HILLSBOROUGH

The Incredible World of Insects

Incredible Ants

By Susan Ashley

Please visit our website, www.garethstevens.com. For a free color catalog of all our high-quality books, call toll free 1-800-542-2595 or fax 1-877-542-2596.

Library of Congress Cataloging-in-Publication Data

Ashley, Susan.
Incredible ants / Susan Ashley.
   p. cm. — (The incredible world of insects)
Includes index.
ISBN 978-1-4339-4572-4 (pbk.)
ISBN 978-1-4339-4573-1 (6-pack)
ISBN 978-1-4339-4571-7 (library binding)
1. Ants—Juvenile literature. I. Title.
QL568.F7A743 2012
595.79'6—dc22

2010035219

New edition published 2012 by
Gareth Stevens Publishing
111 East 14th Street, Suite 349
New York, NY 10003

New text and images this edition copyright © 2012 Gareth Stevens Publishing

Original edition published 2004 by Weekly Reader® Books
An imprint of Gareth Stevens Publishing
Original edition text and images copyright © 2004 Gareth Stevens Publishing

Designer: Daniel Hosek
Editors: Mary Ann Hoffman and Kristen Rajczak

Photo credits: Cover, pp. 1, 11, 15, 17, 19 Shutterstock.com; pp. 5, 9 © Robert & Linda Mitchell; p. 7 © Tammy Gruenwald/Weekly Reader Early Learning Library; p. 13 © Scott Camazine; p. 21 Last Refuge/Getty Images.

All rights reserved. No part of this book may be reproduced in any form without permission in writing from the publisher, except by a reviewer.

Printed in the United States of America

CPSIA compliance information: Batch #CS11GS: For further information contact Gareth Stevens, New York, New York at 1-800-542-2595.

Contents

An Ant Colony 4
The Queen 8
Drones 10
Worker Ants 12
How Do Ants Look? 14
Glossary 22
For More Information 23
Index 24

Boldface words appear in the glossary.

An Ant Colony

Ants live in communities. An ant community is called a colony. The ants live and work together. A colony can have a few ants or thousands of them.

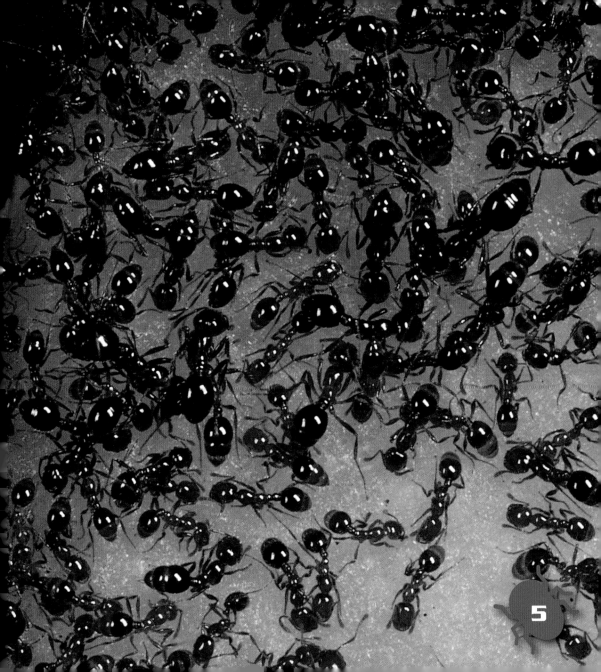

Colonies are above the ground and below it, inside logs, or even inside trees! A colony is called a nest. It has many rooms. The eggs are in one room. The **larvae** and **pupae** grow in other rooms. Tunnels connect the rooms.

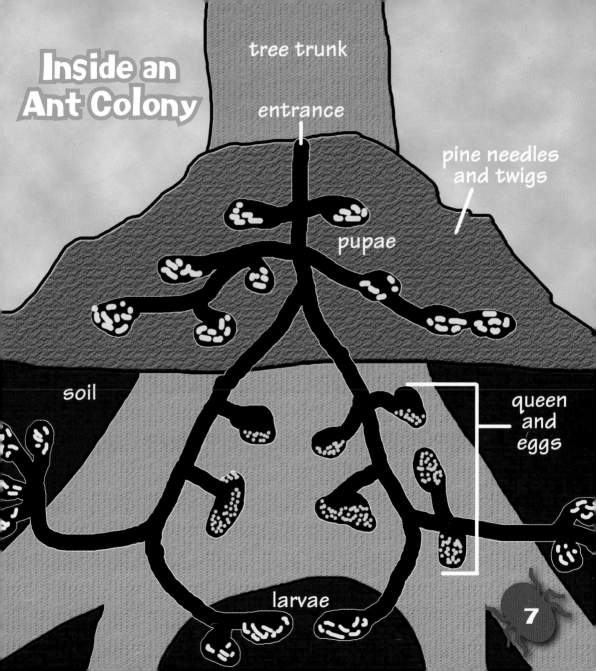

The Queen

Each ant colony has a queen. Some have more than one! The queen is the largest ant in the colony. She spends all her time laying eggs.

Drones

Drones are male ants. They have wings. Their only job is to **mate** with the queen. After that, the drones die.

Worker Ants

Most ants in a colony are worker ants. Worker ants are females, but they do not lay eggs. They take care of young ants. They look for food. They clean and guard the colony.

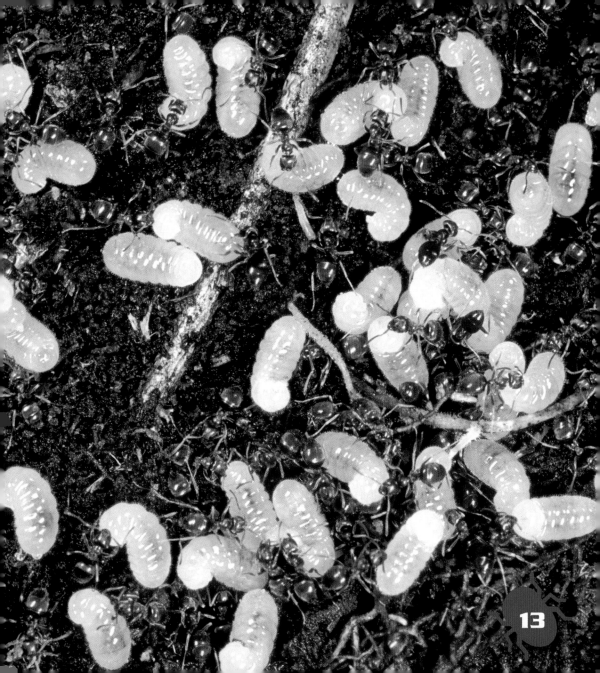

How Do Ants Look?

Ants have six legs. They have a claw on the end of each leg. The claws help them climb and hang. They can even walk upside down!

Ants have two **antennae**. They use them to taste, touch, and smell. They use them to find food. They also "talk" to each other by touching their antennae.

antennae

17

Ants have two very strong **jaws**. The jaws open and close sideways—like a pair of **scissors**. Ants carry things with their jaws. They can lift things that weigh 20 times more than they do!

Ants also have a **skeleton** on the outside of their bodies. They breathe through tiny holes in their skeleton.

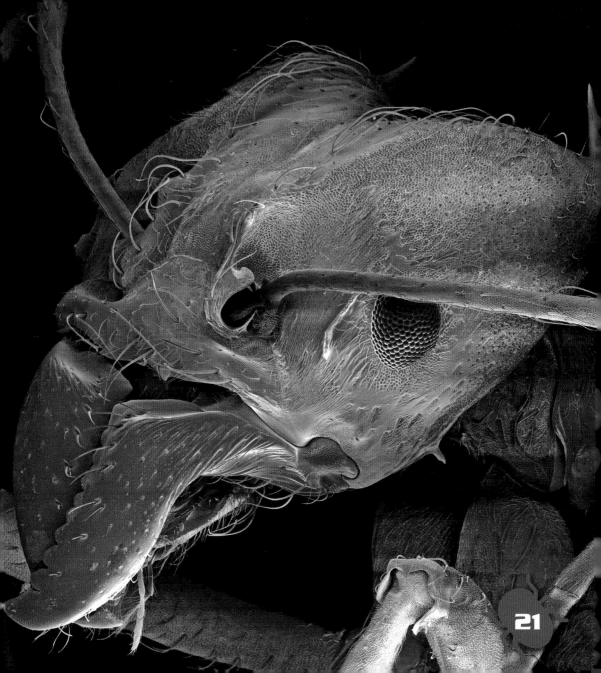

Glossary

antenna: a feeler on the head of an insect. The plural is "antennae."

jaws: mouth

larva: the wormlike second stage of growth of some insects. The plural is "larvae."

mate: to come together to make babies

pupa: the third stage of growth of some insects. The plural is "pupae."

scissors: a tool used for cutting

skeleton: the matter that supports the body of an animal

For More Information

Books
Micucci, Charles. *The Life and Times of the Ant*. New York, NY: Houghton Mifflin, 2006.

Stewart, Melissa. *National Geographic Reader: Ants*. Washington, DC: National Geographic Society, 2010.

Web Sites
All About Ants
www.infowest.com/life/aants.htm
Read about and see diagrams of how ants look, live, and work.

Ants
www.greensmiths.com/ants.htm
Learn interesting facts about the different members of an ant colony.

Publisher's note to educators and parents: Our editors have carefully reviewed these websites to ensure that they are suitable for students. Many websites change frequently, however, and we cannot guarantee that a site's future contents will continue to meet our high standards of quality and educational value. Be advised that students should be closely supervised whenever they access the Internet.

Index

antennae, 16, 17
claws 14
colony 4, 6, 8, 12
drones 10
eggs 6, 7, 8, 12
jaws 18
larvae 6, 7
legs 14

mate 10
nest 6
pupae 6, 7
queen 7, 8, 9, 10
rooms 6
skeleton 20
tunnels 6
worker ants 12

J 595.796 ASH

Ashley, Susan.

Incredible ants